Y0-CWW-644

Written by Raphaëlle Brice
Illustrated by Aline Riquier

Specialist Adviser:
James Fitch, Director
The Rice Museum
Georgetown, South Carolina

ISBN 0-944589-30-8
First U.S. Publication 1991 by
Young Discovery Library
217 Main St. • Ossining, NY 10562

©Editions Gallimard, 1984
Translated by Vicki Bogard
English text © Young Discovery Library

YOUNG DISCOVERY LIBRARY

Rice: The Little Grain that Feeds the World

HIEBERT LIBRARY
FRESNO PACIFIC UNIV
FRESNO. CA 93702

Research Library
AIMS Education Foundation
Fresno. CA 93747-8120

YOUNG DISCOVERY LIBRARY

Rice for breakfast, lunch and dinner!

If you lived in China, India, Vietnam, or other places in Asia, you would eat rice every day. A whole bowlful, with meat, fish, or vegetables for supper; just plain for lunch.

How is rice eaten?

It must be cooked, but it can be served hot, cold, salted, or mixed with milk and sugar to make cakes. You can eat crispy puffed rice with milk for breakfast, or crunch on some in a chocolate bar. Little salted rice crackers come in all kinds of fun shapes.
Many soups have rice in them.

8

Rice grows best with its feet in water and its head in sunlight.

Its stalks are hollow like straws. They draw water up from the roots to the leaves, feeding the whole plant.

wheat

corn

oats

Rice belongs to the grass family. It is a cereal grain like corn, wheat and oats. It can be cooked whole or ground into flour.

We eat only the kernel part of the rice plant.

Look at this grain of rice cut in half. It is protected by an envelope: the **husk.** The kernel contains a seed, which if planted would grow. The plants are leafy and grow panicles, or heads. Each panicle holds 60 to 150 kernels of rice.

Some rice is long grain, and some is short grain.

Rice has been grown for a very long time.

The people of China and India  knew how to grow it 7,000 years ago. Rice growing has not changed much since ancient times.

Rice is a world-traveler.

Merchants, warriors and navigators carried rice from Asia to Africa and other places. Moors from northwestern Africa brought rice to Europe only 800 years ago.

How did rice first get to America?

About 350 years ago, a ship sailing from Africa was wrecked off the coast of South Carolina. The captain gave sacks of rice to his rescuers.

Today, over half the world eats rice as its main food. Nine out of ten sacks of rice grown in the world come from Asia. Asian people keep most of that rice for themselves.

American states that grow rice
include Texas, Louisiana
and California, but the largest
crop is in Arkansas.
Rice is exported far around
the world from the United States.

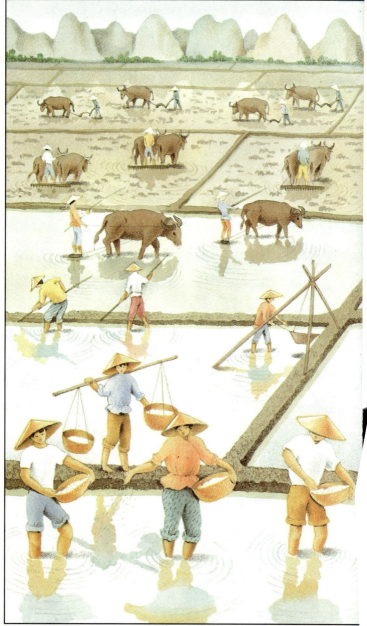

Water buffalos move easily in the muddy rice paddies. They do tire quickly and can only work for about two hours a day.

This is Asia, the land of rice. To grow rice, farmers turn their

A tractor with wheel-cages to support it in the mud.

fields into **rice paddies.** First they build low dirt walls, called dikes, around the fields. Irrigation canals bring water from a river to flood the fields. Then the ground is leveled and the soil is worked into a soft mud. This is called puddling. Now the paddy is ready, but the rice seeds must first be sown in a separate field: the seedbed.

Sometimes water is taken from the paddy to water the seedlings.

The first tillers, or shoots, appear 5-10 days after planting. After a month, they will be too crowded. Time to move to the paddy!

Women transplant the rice, backs bent, their feet in the mud.
The rice shoots are replanted in rows with plenty of room to grow. The water keeps the weeds from growing too high. It takes three to six months for the rice to ripen. While they wait, the farmers keep the water at the right level and chase birds and bugs away.

Did you know that there are thousands of different kinds of rice in the world?

The rice paddy is bird heaven.

Long-legged herons, cranes
and wild ducks find all their
favorite foods in a paddy:
frogs, fish, snails,
weeds…and grains
of rice.

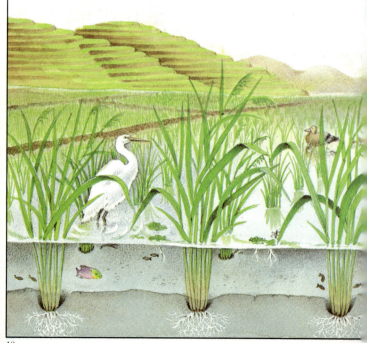

The wet fields also attract mosquitos, which can carry diseases. Some insects, like stem borers and leaf hoppers, love to devour the stalks. Others eat the rice itself.

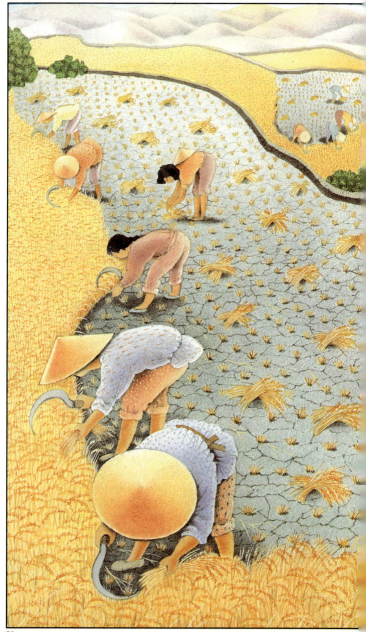

The rice plants turn golden and are full of grains.

The rice dries out while it finishes ripening.
Just before harvest time, the farmers let the water flow back to the river. The rice paddies dry out, turning to hard mud.

Harvest time.
From dawn until dusk, the rice growers cut the stalks with sickles, then tie them into bundles. They carry them in wheelbarrows or on their backs to the village. What happens to the rice then?

Threshing the rice.

The farmers beat the stalks against a big stone roller to separate the grains from the rest of the plant. They make haystacks out of the stalks. The grains are spread out and raked to dry in the sun.

The rice is hulled.

To remove the hulls from the grains, the rice is put in a mortar. With a pestle, the grains are crushed and rubbed against each other. The hulls burst open and release the kernels.

The rice is winnowed.

The kernels are put on big bamboo mats and shaken. The light hulls blow away and the heavier grain is left.

A rice loft

In Asia, rice is even grown in the mountains. Terraces, like steps, are dug on the slopes. Stone walls around the terraces hold the water in the paddies.

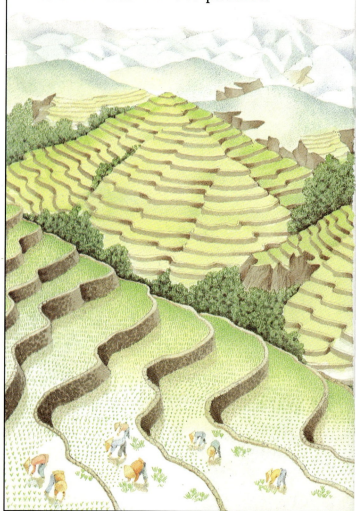

Water for the rice comes from rainfall and from mountain streams. The water flows down from terrace to terrace through little canals.

The land of machines.
In the United States, rice is
grown in enormous fields.

Almost all the work is done by
big, powerful machines.

The soil is plowed and tilled using
tractors. There is even a machine
to build the dikes, and pipelines to
flood the fields.

First the seeds
are **sprouted** in
wet sacks.

Low-flying planes scatter the
sprouted seeds.

Machines, called **combines**, harvest
and thresh the rice. At the
mill, the rice is dried, hulled
and packaged.

Is rice only good for eating?

Rice straw is made into handbags and sleeping mats and paper. The straw is used for those sun hats the rice farmers wear.

The rice hulls can be fed to livestock, the buffalos or pigs. Hulls are also used as fertilizer. They are spread on the fields to enrich the soil for the crop next year.

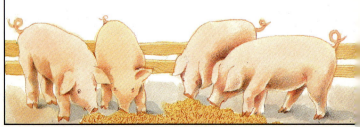

Asian people grind broken rice grains into flour for cakes and noodles. They also make beer, and wine called **sake**, from rice. Women use rice powder as makeup. By adding sugar rice candy is made.

White rice or brown?

With the hulls still on, it's called paddy rice. Under the hull are **bran** layers that contain most of the vitamins and minerals in the kernel. Hulled rice is called brown rice. If you want white rice you must use special machines to remove the bran layers. Most people eat white rice.

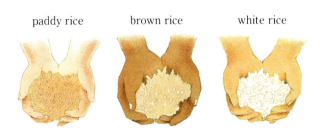

paddy rice   brown rice   white rice

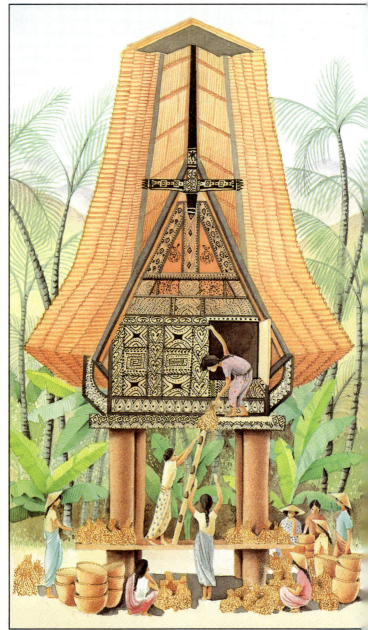

Rice Festivals

In Asia, people have festivals to ask the gods for a good harvest. Each year, in the mountains on the

Indonesian island of Sulawesi, the Toradjas sing all night and have dances that mimic their labor in the rice paddies.

Side view of a Toradja rice loft.

Every family has a rice loft next to their house. It is built on stilts and has a curved roof to protect the precious grain from rain and rising ground water.

The people of Bali offer rice cakes to the temple gods.

Congratulations!

In the United States and Europe, we throw rice at newlyweds to wish them health, wealth and happiness.

How is rice cooked?

In Asia it is usually steamed in a basket made of bamboo, set on a pot of boiling water.

This Jamaican woman is cooking rice in a big pot of salted boiling water. In Italy and Turkey, rice is cooked in a big skillet with pieces of meat, vegetables and spices.

In countries where rice is eaten every day, people do not get tired of it. It is like you having bread with most of your meals.

Brown or enriched white rice has vitamins for healthy skin and hair.

To serve 4, you need:

2 cups rice

3 eggs

butter or oil

a half-pound of lean pork

cooked peas

salt & pepper

Cantonese-style fried rice for the whole family.

Cook white or brown rice (follow package directions). Cut the pork into small cubes and cook well in light oil. Beat eggs for an omelet, cook and cut it in strips. Lightly sauté the rice in an oiled pan, stirring often. Lay rice, then peas, pork and egg strips on a plate. Salt, pepper and dot with butter.

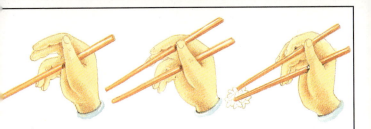

How do you eat with chopsticks?
Do as the Chinese and Japanese
do: use a short grain rice that
is a little sticky. It's easier
to pick up. Study the pictures.
Hold one stick between your thumb
and ring finger. That stick does
not move. Hold the other one
between the tip of your thumb and
your middle finger. Use that
stick to pick up food by pressing
it against the other stick.

Rice Pudding

What is the matter with Mary Jane?
She's perfectly well, and she
   hasn't a pain.
And it's lovely rice pudding
   for dinner *again*,
What *is* the matter with Mary Jane?

A.A. Milne

With coarse rice to eat,
  with water to drink,
and my bended arm for a
  pillow—
I still have joy in the
  midst of these things.

Confucius

HIEBERT LIBRARY

3 6877 00226 7093

Index